Workbench
SILENCERS

WARNING

It is against the law to manufacture, purchase, or possess a firearm silencer without proper authorization from and registration with the Bureau of Alcohol, Tobacco, and Firearms of the U.S. Treasury Department. Inquiries regarding the procedures for applying for a license to construct, purchase, or possess a silencer and for paying the appropriate tax must be directed to that office.

In addition, many state and local jurisdictions have restrictions on the possession of silencers, even if appropriately taxed and registered with federal authorities.

Do not attempt to construct a silencer without proper authorization from federal, state, and local authorities. This book is presented for academic study only.

• • • • •

The technical data presented here, particularly concerning the use, adjustment, and alteration of firearms, inevitably reflects the author's beliefs and research with particular firearms, equipment, components, and techniques under specific circumstances which the reader cannot duplicate exactly. The information in this book should therefore be used for guidance only. Neither the author, publisher, or distributors of this book assume any responsibility for the use or misuse of information contained herein.

Workbench
SILENCERS

The Art of Improvised Designs

George M. Hollenback

Paladin Press · Boulder, Colorado

Also by George M. Hollenback:

More Workbench Silencers

To Mom

Workbench Silencers:
The Art of Improvised Designs
by George M. Hollenback

Copyright © 1996 by George M. Hollenback
ISBN 0-87364-895-1
Printed in the United States of America

Published by Paladin Press, a division of
Paladin Enterprises, Inc., P.O. Box 1307,
Boulder, Colorado 80306, USA.
(303) 443-7250

Direct inquiries and/or orders to the above address.

Visit our Web site at www.paladin-press.com

Contents

Apology

No, I'm not about to say that I'm sorry for anything. This is going to be an apology in the classic sense of the word—you know, like the *Apology* of Socrates (or So-crates as some call him). I'm going to explain my reasons for writing this book.

I am a part-time inventor who became interested in firearm silencers. I read up on the subject, researched prior patents, and came up with what I believed to be a new design.

Before I could begin work on a prototype, however, I had to obtain approval from the Bureau of Alcohol, Tobacco, and Firearms (BATF). Manufacturing a firearm silencer without BATF approval is a *federal offense*. In fact, putting *anything* on the end of a gun to reduce its noise level is a federal offense, even if it's just a pop can or plastic soft drink bottle.

I obtained an "Application to Make and Register a Firearm" form from the local branch of the BATF and filled it out. The form also required my photo, fingerprints, and the signature of the chief law enforcement officer in my area.

I contacted the ID division of my local police department, who agreed to fingerprint me, run a Brady Bill check on me, and forward my application to the chief of police for his signature. They also informed me that their own procedures required a letter of introduction to the chief of police and two photos of a secure storage area for the silencer.

I typed a letter for the chief explaining that I was an inventor wanting to work on a new silencer design. I even enclosed a photocopy of a patent I had just received on my first invention, an exercise device, just to show that I was indeed a legitimate inventor. I further explained that I had resided at my current address for more than 10 years, had been employed by the same company (a major hospital) for more than 10 years, and that I had a clean police record. I purchased a gun vault and photographed it *in situ* at my apartment.

When I reported for fingerprinting with my assembled application materials, ID personnel informed me that it would probably take at least three weeks for my application to be processed. They would send it back to me, and I in turn would send it on to the BATF.

During the time I had been doing my research, I had come across many examples of improvised silencers. Some of these devices were so easily and cheaply fabricated that they were dubbed "disposable silencers." My inventor's mind became so fascinated with this concept that I couldn't walk into a hardware store or builders supply store without stopping to examine sundry items to see if they could possibly be assembled together as silencer components.

Much to my surprise, I discovered that a great variety of disparate objects harmoniously lent themselves to that end. Some of my discoveries were so simple and elegant that I wondered why someone else hadn't thought of them before. I continued to amuse myself with this little minimalist engineering game during the three and a half weeks I waited to hear from the police department.

When the big envelope finally arrived, I opened it with the expectation of finding a signed application that I could immediately forward to the BATF. Well, the chief of police had signed the application—right on top of a big rubber stamp that read DISAPPROVED.

No, they hadn't found anything in their background check that would disqualify me from manufacturing and owning a silencer. As I stated earlier, my record is squeaky clean. It's just that chief law enforcement officers are under no obligation whatsoever to sign the form. It was probably a C-Y-A move on the chief's part: what if he signed this form for someone, and that person later went berserk, screwed his silencer onto a gun, and went out and popped someone?

It still made me angry, though. Research had revealed that my city had a reputation as a center for illicit silencer manufacture. Underground workshops were turning out silencers for crooks, while a law-abiding inventor playing by the rules was being denied permission to work on a silencer legally!

I hated to think that all my research had been for naught. Was there anything I could salvage from my aborted patent venture?

Well, I thought, what if I compiled my informal research on improvised silencers into a book? If I couldn't get inventor's royalties, I might as well try for author's royalties.

This presented an interesting technical problem, however. For the sake of realism, I wanted the work to be illustrated by photos rather than drawings. (A drawing of an improvised silencer sometimes leaves you wondering if the thing was ever built at all; a photo leaves no doubt). But how could I build and photograph these silencers without breaking federal law?

I elected to stage an elaborate illusion. I could show how certain common items would fit together but stop short of actually producing a complete silencer. The finished product shown on the end of a gun would actually be a plaster of paris filled dummy. Couplings and adapters would have ball bearings epoxied into their throats, making them useless as real silencer accouterments. There's no law against manufacturing realistic-looking props and photographing them in such a way that they look like the real thing.

But won't a book like this enable unlicensed people—even criminals—to manufacture illegal silencers?

First of all, improvised silencer technology has been around for a long time. Any criminal bent on making and using an improvised silencer will indeed do so and would have done so had this book never been written.

Second, for all we know, the ideas in this book may already be in use by criminals. Gottfried Wilhelm Leibniz and Sir Isaac Newton, working independently of each other, both came up with the mathematical tool called calculus. Now if great minds like Leibniz and Newton can independently invent something like calculus, it certainly stands to reason that lesser minds can independently invent something like an improvised silencer. Unlike me, however, a criminal who comes up with one of these improvised silencer designs is not going to publish it or scrupulously adhere to BATF rules about making it. On the contrary, he's going to manufacture it illegally, use it for whatever nefarious purpose he has in mind, then keep it to himself or share it only with other criminals.

For this reason, this book is a valuable reference work that belongs in the library of everyone involved in forensic science, law enforcement, or private investigation. By understanding just how amazingly easy it is to fabricate silencers from readily available materials, these professionals can be alert for evidence they might otherwise overlook.

Not everyone who builds an illegal improvised silencer is a real criminal, though. There are some otherwise law-abiding citizens out there who, out of insatiable curiosity, will build and try these designs. Yes, technically this is a federal crime. But let's be realistic: some Walter Mitty farting around in his basement with an improvised silencer somewhere is no threat to you, me, or anyone else except maybe himself.

Licensed silencer manufacturers will also be among those who try these designs. Many of these pros have sophisticated sound measuring instruments with which they can compare the performance of different silencers. Sometimes they find that an improvised unit performs unexpectedly well when compared with expensive machined units. (I am curious how my designs will stack up against the competition. Send your research findings and any accompanying praise or excoriation to me in care of the publisher.)

Finally, there is that great majority of readers out there who will never ever actually construct an improvised silencer. We've already mentioned the professional crime fighters who will use this as a reference work. There are also professional and aspiring writers who like to do their "homework" on things like silencers before composing works dealing with crime and espionage. (Would God that more of them, particularly scriptwriters, did this kind of homework!) Some folks are connoisseurs of arcane knowledge relating to weapons, munitions, James Bond gadgetry, and the like. Although they don't actually put this knowledge to use, they take smug delight in simply knowing that they possess it. (These are the people who like to pick mistakes out of movies, books, and TV programs made by writers who didn't do their homework.) Most readers, though, are going to be fairly ordinary folks with a healthy curiosity who like to read about subjects a little off the beaten path.

So, for whatever it's worth, here's my little contribution to that fascinating field known as silencerology.

Introduction

The word "silencer" is somewhat of a misnomer; there's always going to be some kind of noise associated with the discharge of a firearm. Laymen who watch videos showing silencers in action are sometimes dismayed by the racket produced by supposedly "silenced" weapons.

This disillusionment is the result of the influence of Hollywood on popular culture. Who among us has not seen on TV or in a movie a large-caliber handgun or even a high-power rifle rendered whisper quiet by the installation of a small cylindrical device on the end of the weapon's barrel?

What a silencer really does is *suppress* the level of sound produced by a firearm. If the loud *bang* produced by a Colt .45 can be reduced to the mild *pop* of a .22 Short, the noise level of the weapon can be said to have been effectively suppressed. If the weapon is a small caliber to begin with, the noise level of the muzzle blast may be so suppressed that it is quieter than the mechanical workings of the gun itself. A friend once witnessed a demonstration of a silenced .22 semiautomatic pistol. He said the only noise he heard was the clack of the bolt moving back and forth, and then, a split second later, the slap of the bullet hitting the target. Suppression this efficient is about the closest thing there is to "silencing" a gun.

There are four major sources of firearm noise that need to be understood in relation to silencing or suppressing a weapon:

MUZZLE BLAST

When a bullet leaves the barrel of a gun, it is followed by a mushrooming cloud of hot, burning, expanding gases. When these gases collide with the cooler surrounding air, a loud blast is produced. If these hot gases can be contained and delayed for just a split second, it slows them down and cools them down just enough that they make much less racket when they finally escape into the atmosphere. The silencer is the device that regulates the release of these gases. It may embody any of the following features:

1) *Expansion Chamber*—An enclosed space that briefly contains the gases before they follow the bullet out of the silencer.
2) *Baffles*—A series of partitions through which the bullet must pass. The simplest baffles are just washers mounted at intervals in a tube. Baffles split the silencer up into a bunch of little expansion chambers.
3) *Perforated Tube*—A slotted or drilled tube that makes up the core of some silencers. The gases following the bullet down the tube are vented out through the perforations and absorbed by diffusing material surrounding the tube.
4) *Wipe*—A relatively thick rubber disk penetrated by the bullet. The gases following the bullet are sealed off behind it as the bullet burrows its way through the wipe.
5) *Packed Tube*—Wire mesh washers or the like packed solid in the main silencer tube. The gases following the bullet are forced into the tiny spaces in the packed material.

Nearly all the silencers in this book work on the "expansion chamber" principle because rigging a hollow container onto the end of a gun is such a relatively simple procedure. These containers, however, may be lined, stuffed, or covered with sound-dampening materials to enhance their efficacy.

A number of designs use wipes made from a variety of objects. One design, a sponge stuffed into a plastic shell, is actually nothing but one big wipe.

A couple of designs involving containers arranged end to end incorporate the baffle principle: the bullet must pass through a little hole in the end of one container in order to enter the adjoining container.

Three designs show silencer tubes that can easily accommodate simple baffles or packed washers.

The only basic silencer principle not shown or suggested in this book is the perforated tube. It requires a lot of slotting or drilling, very close tolerances, and near perfect alignment that is very difficult to achieve.

BREECH BLAST

With certain guns, not all the gases exit the muzzle; some of the gases end up escaping from around the rear of the barrel. When these gases collide with the cooler surrounding air, they too produce noise. Putting a silencer on the front of the barrel won't suppress noise coming from the rear of the barrel.

Revolvers have a gap between the cylinder and the barrel that allows gases to escape. The whisper-quiet silenced revolvers seen on TV and in the movies are a myth; the blast from the cylinder gap would make quite a racket. (Unless, of course, the revolver were an M1895 Nagant; this unique weapon incorporates mechanical features that seal the gap between the cylinder and the barrel every time the hammer is cocked.)

Automatic and semiautomatic weapons have bolts or slides that fly back with each shot, extract and eject the spent cartridge, then slam forward and chamber the next round. If there are burning gases still in the barrel when the breech opens up during this process, these gases will escape from the rear of the barrel.

Making sure the breech stays closed eliminates this problem. Special slide locks can be installed to accomplish this, or the shooter can simply brace the heel of his hand against the back of the slide or bolt. Another strategy is to use low-power ammunition that doesn't have the energy to push the slide or bolt back. (Low-power ammunition is quieter, too, so there's less noise to silence.)

MECHANICAL NOISE

The interaction of the moving parts of a gun produces noise apart from the actual sound of the weapon's discharge. For example, just racking back the slide or bolt of an automatic or semiautomatic weapon and letting it slam forward can make quite a bit of clatter in and of itself. When the weapon cycles during firing, that noise is still going to be there, regardless of how efficient a silencer might be screwed onto the barrel. The problem is even more acute in fully automatic weapons, where a steel bolt in a steel receiver might slam back and forth a dozen or more times in a single second.

Eliminating the slamming of the bolt or slide can be accomplished by the same remedies for breech blast: manually holding the bolt or slide down, mechanically locking the slide, or using low-power ammunition.

Sometimes, as with the Colt .45, the weight of a heavy silencer on the barrel can eliminate the movement of the slide. This occurs because of the way the barrel and the slide work together when the weapon is fired. Ordinarily, the recoil that pushes the slide back also pulls the barrel back a short distance because the slide is locked onto the barrel by several interlocking grooves. The back of the barrel then cams slightly downward, disengaging these interlocking grooves, and the freed slide flies all the way back. (Keeping the barrel and slide locked together for a short time gives the bullet time to clear the barrel before the breech opens up, thus eliminating or minimizing breech blast.) When a heavy enough silencer is hung on the end of the barrel, the slide can't pull the barrel back far enough for the interlocking grooves to disengage and free the slide.

SONIC CRACK

When a jet aircraft tears through the air faster than the speed of sound, it creates a peculiar kind of noise that radiates out behind it in a conical pattern. People see the plane fly overhead, then later hear a "sonic boom" as the outer fringe of this giant invisible cone drags over the surface of the earth where they're standing.

The same thing happens on a smaller scale when a bullet travels faster than the speed of sound (about 1,100 feet per second, or fps). People standing downrange hear nothing as the bullet zips by; they then hear the sonic crack of the bullet, followed by the distant sound of the gunshot itself. Even if the muzzle blast of the gunshot is silenced, the sonic crack will still remain if the bullet is traveling faster than the speed of sound.

The easiest solution to the sonic crack problem is to use subsonic ammunition with a muzzle velocity rated lower than 1,100 fps. Some ammunition, such as .22 Short or .45 ACP usually falls well below the 1,100 fps sonic threshold. Other ammunition, such as .22 Long Rifle, is available in both supersonic and subsonic factory loads. Ammunition that is invariably supersonic, such as that fired by most centerfire rifles, usually must be specially loaded to subsonic velocity.

Supersonic ammunition can sometime be rendered subsonic by the silencer itself. Silencers constructed around perforated gun barrels absorb gases that are vented out through the perforations in the barrel. If enough gases are bled off this way, the velocity of the bullet can be reduced substantially. The velocity of a bullet can also be reduced if the bullet has to burrow through a series of wipes on its way out of the silencer.

Sometimes, though, lowering the velocity of the ammunition to avoid sonic crack just isn't feasible. In long-range sniping, for example, lowering the velocity of the bullet would rob it of the power it needs to reach and penetrate the target. In situations like this, simply suppressing the muzzle blast is all that can be done to make the sniper less conspicuous. Of course, if the sniper is far enough away, the velocity of the bullet may drop to subsonic levels before it reaches the target.

Tools and Materials

Nearly all the designs in this book require no more tools than a hacksaw, screwdriver, kitchen knife, modeling knife, scissors, can opener, and metal punch (or big nail). There are a few exceptions: one design requires a drill, one requires a chisel blade for the modeling knife, one requires a socket wrench, and one requires a means for cutting longitudinal slots in a short length of PVC tubing. (I used a hand-held Dremel Moto-Tool to do this. It could probably be done with a hand saw if the section of tubing is held in a vise.)

Adhesives—tapes and glues—are also used in many of these designs. Tapes used include masking, black vinyl, and metal repair tape in widths ranging from 3/4 inch to 2 inches. Glues used include epoxy, PVC cement, and anything suitable for permanent rubber-to-rubber and rubber-to-plastic bonds (like Goop). Liquid Steel and silicone sealant were used for a couple of fill applications.

The tapes are mostly used for "friction fitting"—wrapping just enough tape around one cylindrical object so that it will fit very tightly inside another cylindrical object. It takes a little practice to get it right: a mere half or quarter of a wrap might make the difference between a fit that's too loose or too tight and one that's just right. You may have to wrap, unwrap, and snip the tape several times. For added strength, wrap tape around the outside of the joint formed by the friction-fit components. If you don't plan on disassembling the silencer, you can go ahead and glue the friction fittings together.

Some components, however, may already fit together so snugly that wrapping even a single layer of tape around one will make it too wide to fit back inside the other. In cases like this, you'll need to use paint instead of tape. Spray a coat of primer on the component, let it dry, and try the fit. Repeat until a tight enough fit is obtained.

Some designs involve squeezing PVC bushings into snug-fitting rubber components. If the fit is too snug, you'll need to have some water-soluble gel on hand to use as a lubricant.

And finally, a variant of one design calls for the application of some lithium-based grease.

In order to avoid repetition, the tools and materials listed above won't be relisted along with the components that make up each silencer. We'll simply assume that you already have them all on hand.

Weapon #1
Ruger Bull Barrel Pistol

This is probably one of the most suitable weapons there is for improvised silencers. It has a reputation as a rugged, dependable gun and is such a common item that finding one for sale or ordering one is no problem. Also, .22 caliber weapons are the most easily silenced of firearms.

The best feature of this weapon, though, is the ease with which the silencer can be attached to the barrel. The bull barrel is a long, straight cylinder with no taper. The front sight can be removed by taking out a single screw. Rigging a silencer onto a plain cylindrical barrel is much easier than rigging one onto a tapering barrel with a fixed front sight. Also, keeping the silencer aligned with the barrel is much easier when the barrel has a simple cylindrical shape.

COUPLING #1

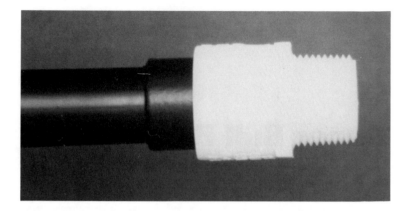

In silencer parlance, a coupling is a device that connects the silencer to the gun barrel. The simple coupling shown here on the Ruger bull barrel is created by removing the front sight, wrapping the end of the barrel with 1 1/2 inch vinyl tape, then friction fitting a 3/4 inch PVC male adapter tightly over the taped end. This coupling will take any silencer that incorporates a PVC bushing threaded for 3/4 inch stock. It can be used on any cylindrical barrel from which the front sight has been removed, including rifle barrels.

ALUMINUM CAN SILENCERS

Although aluminum beverage cans are so flimsy that you can crush them in your hand, they are remarkably strong for their weight and capable of withstanding a great deal of internal pressure. (Set an unopened can of pop on the floor and stand up on it on one leg. Unless you weigh a quarter of a ton, it ought to support your weight.)

There's nothing new about using aluminum cans as simple, expansion chamber type silencers. What is new is the following series of adapters for holding the cans securely onto the end of the gun. Five of these six adapters use rubber plumbing fittings for standard pipe having an inside diameter of 2 inches. These fittings will take aluminum cans whose tops taper to a 2 1/4 inch diameter rim. (The smaller rims on some cans don't provide a tight enough fit.) The sixth adapter is a scaled down version of one of the others, using a smaller rubber fitting and taking a V-8 tomato juice can instead of a standard pop or beer can.

A word of warning. One of my research sources states that just an adapter itself—without a can mounted in it—counts as a silencer and must be registered with the BATF.

ALUMINUM CAN #1

Materials needed:
• aluminum can
• 2″ x 1 1/2″ flex coupling
• 1 1/2″ x 3/4″ PVC bushing

Insert bushing into coupling and tighten band clamp.

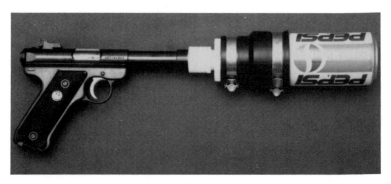

Fit adapter onto top of can, tighten other band clamp, and install on weapon.

ALUMINUM CAN #2

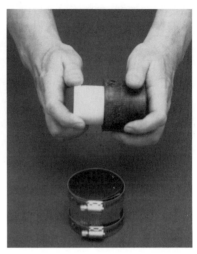

Materials needed:
- aluminum can
- 2″ x 1 1/2″ no-hub coupling
- 1 1/2″ x 3/4″ PVC bushing

Loosen band clamps and remove metal sleeve from coupling. (The band clamps on this particular coupling had to be loosened with a socket wrench instead of a screwdriver.) Insert bushing.

Fit adapter onto top of can, replace sleeve, and tighten band clamps.

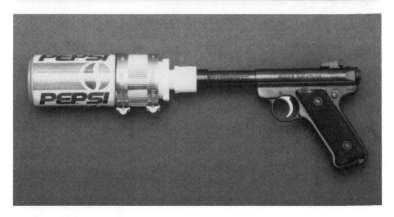

Install on weapon.

ALUMINUM CAN #3

Materials needed:
- aluminum can
- drain trap connector (large)
- 1 1/4″ x 3/4″ PVC bushing
  (no taper inside)
- 1″ x 3/4″ PVC bushing
  (unthreaded)

Insert 1 1/4″ x 3/4″ bushing into connector.

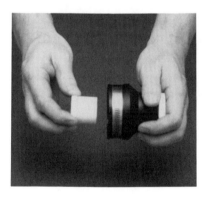

Glue smaller bushing into larger bushing as shown. If inside walls of larger bushing are straight instead of tapered, smaller bushing should easily fit inside.

Build up layers of silicone sealant inside adapter until flush with top of smaller bushing. Let each layer harden before applying the next. Break off the end of a Popsicle stick, wet it, and sweep it over the last layer of sealant to smooth it out before it dries.

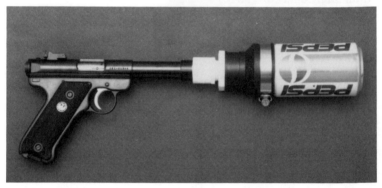

Fit adapter onto top of can, tighten band clamp, and install on weapon.

ALUMINUM CAN #4

Materials needed:
- aluminum can
- 2″ pipe cap
- 1″ x 3/4″ PVC bushing
- basin gasket

Cut hole in cap as shown. (Use the bushing to trace a circle in the center of the top of the cap. Make short, plunging cuts with a modeling knife blade angled slightly inward.)

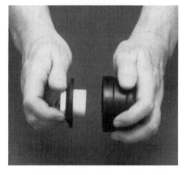

Insert bushing into basin gasket.

Glue bushing/basin gasket assembly onto top of cap.

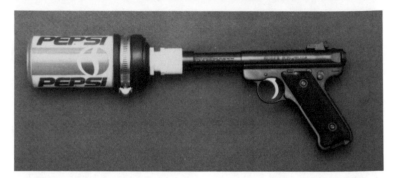

Fit adapter onto top of can, tighten band clamp, and install on weapon.

ALUMINUM CAN #5

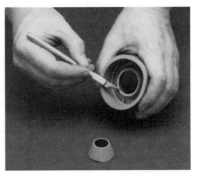

Materials needed:
• aluminum can
• drain hose connector
• 1″ x 3/4″ PVC bushing
• basin gasket

Cut off portion of funnel-shaped seal as shown. (Turn it inside out, then cut around it with modeling knife.)

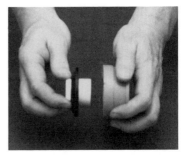

Insert bushing into basin gasket.

Glue bushing/basin gasket assembly onto top of connector.

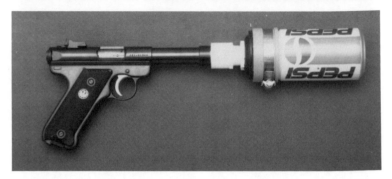

Fit adapter onto top of can, tighten band clamp, and install on weapon.

ALUMINUM CAN #6

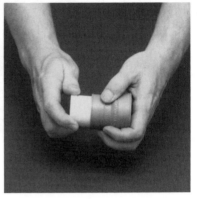

Materials needed:
- V-8 cans (2)
- drain trap connector (small)
- 1 1/4″ x 3/4″ PVC bushing

Insert bushing into connector.

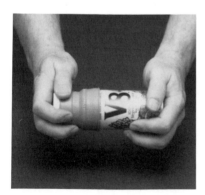

Fit adapter onto top of can.

Tape second can onto first with metal repair tape.

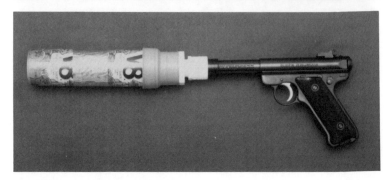

Install on weapon.

ALUMINUM CAN PREPARATION

Bare cans can be used with little preparation. Rotate the pull tabs 180 degrees and pull straight up on them; they should pop right off. Then carefully insert the tip of your little finger into the can and push the flap back as far as it will go. Unless you're a real klutz, you ought to be able to do this without cutting yourself.

A simple bare can can easily be bolstered from the outside with a drink koozie and a wipe.

 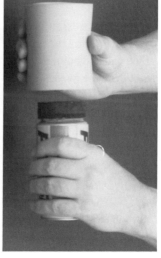

Materials needed:
• aluminum can
• drink koozie
• 2 1/4″ foam rubber disk

Use the top of the can to trace a circle on one of those oblong foam rubber wrist rests used with CRT keyboards. Cut it out with scissors. Turn the can upside down, place the disk on the bottom of the can, and push the koozie down over disk and can.

Install on weapon.

Another design in-
volves removing the top
of the can and augment-
ing it from the inside.

Materials needed:
• aluminum can
• steel wool pads
  (two medium-
  to coarse-grade pads)
• 3/8" dowel (sharpened
  on one end)

Cut out top of can. The type of can
opener shown here works perfectly.
(Don't cut in the groove running
around rim; cut further in. Run the han-
dle of the can opener around inside of
rim to smooth back jagged edges.)

Punch 1/4″ hole in center of can bottom.

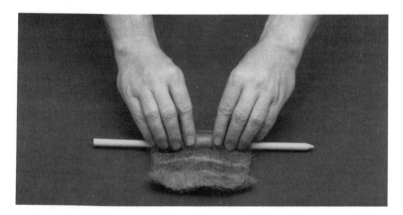

Unroll steel wool pads and reroll them around dowel. Use medium or coarser grades; finer grades are crumbly and flammable. (Remember the old Boy Scout stunt of starting a fire by rubbing a piece of flint against a file and letting the sparks drop onto a ball of fine steel wool?)

Insert pointed end of dowel into hole in bottom of can. Keeping the dowel centered and fixed in this position, work steel wool into can. Then remove dowel; this will leave a channel in steel wool running length of can.

(There is a possibility that the force of the muzzle blast might fluff up the steel wool enough to occlude the channel made by the dowel. You can give the steel wool a little more cohesion by squirting several beads of lithium-based grease lengthwise and crosswise on the unrolled pads and coating the inside of the can with a generous layer of the same. You can also try using a larger dowel to increase the diameter of the channel.)

Smaller V-8 cans can be packed with steel wool just like pop cans. Use only one steel wool pad per can.

V-8 cans can also be lined with fiberglass insulation for 1/2 inch pipe. Cut the side of a toilet paper roll tube and use it to "shoehorn" the segment of insulation into the can. (Be careful when handling the fiberglass; it can irritate the skin and stick to your fingers.)

BUBBLER SILENCER

Materials needed:
- bubbler (also called "soaker")
- natural sponge (trimmed to size and shape shown)

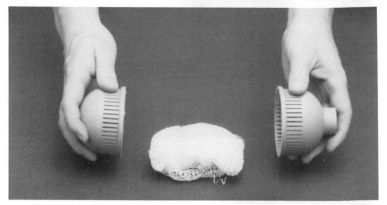

Bubblers are attached to garden hoses and used to irrigate shrubs. They have a plastic "pot scrubber" type element inside to diffuse water flow.

Cut or saw bubbler shell in two around seam. Remove and discard diffusing element.

Drill 1/4"-3/8" hole in end of bubbler shell.

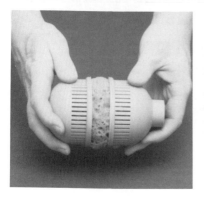 

Wet sponge, wring it out, and fit it into bubbler shell.

Tape halves of shell back together.

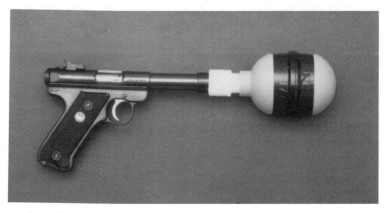

Install on weapon.

CLOG BUSTER SILENCER

Materials needed:
• Clog Buster

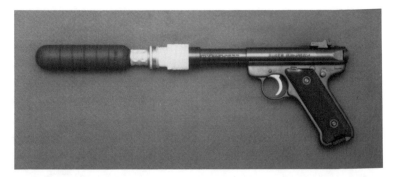

The Clog Buster is a gadget used for unplugging drains. It consists of a tough rubber bladder with a hose attachment at one end and a tiny slit at the other end. No special preparation is needed; just install Clog Buster on weapon.

OIL FILTER SILENCERS

Oil filters consist of a perforated steel sleeve surrounded by a sound-dampening filler element enclosed in a steel shell. Why hasn't anyone thought of using one as a silencer before? The oil filters used with the adapters shown here have diameters of 3 3/4 inches and 3 inches. They have a second layer of sheet metal about 1/4 inch under their top ends, which makes a good double baffle for the bullet to pass through. Look inside the oil filters to make sure there are no brackets or fixtures at the top end that might interfere with the transit of the bullet.

OIL FILTER #1

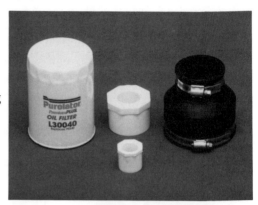

Materials needed:
- oil filter
  (3 3/4″ diameter)
- 3″ x 2″ flex coupling
- 2″ x 1″ PVC
  bushing (un-
  threaded)
- 1″ x 3/4″ PVC
  bushing

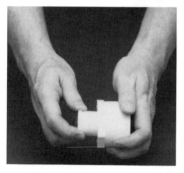

Glue smaller bushing into larger bushing.

Insert bushing into coupling and tighten band clamp.

Make four slits in coupling as shown.

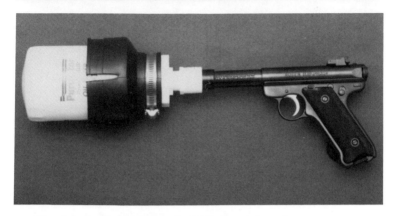

Fit adapter onto bottom of oil filter and install on weapon

OIL FILTER #2

Materials needed:
• oil filter (3 3/4″ diameter)
• 3″ pipe cap
• 1″ x 3/4″ PVC bushing
• basin gasket

Cut hole in rubber cap as shown.

Fit bushing into basin gasket.

Glue bushing/ basin gasket assembly onto top of cap.

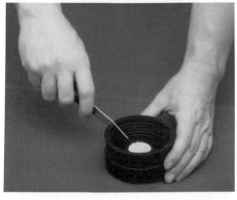

Make four slits in cap as shown.

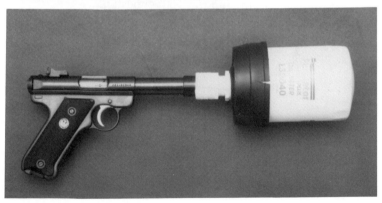

Fit adapter onto bottom of oil filter and install on weapon.

OIL FILTER #3

Materials needed:
- oil filter (3″ diameter)
- section of 3″ cardboard mailing tube
- 2″ x 1″ PVC bushing (unthreaded)
- 1″ x 3/4″ PVC bushing

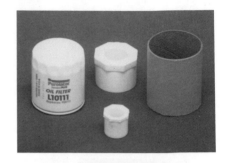

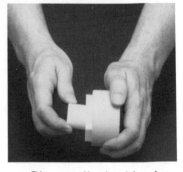

Wrap masking tape around larger bushing, building out to about 3" diameter that will fit tightly into cardboard tube.

Glue smaller bushing into larger bushing.

Friction fit built-up bushing into cardboard tube.

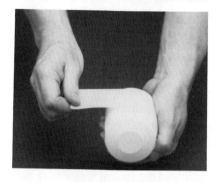

Wrap oil filter with enough layers of tape to ensure tight fit in cardboard tube.

Friction fit oil filter into cardboard tube.

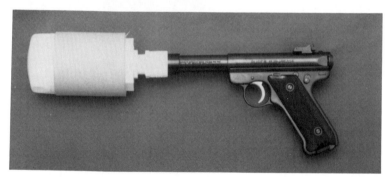

Install on weapon.

PVC SILENCER TUBES

Many silencer plans call for metal tubes with machined metal end caps, which can only be fabricated on expensive machine shop equipment. A poor man's version of a silencer tube with end caps can be fabricated in less than five minutes from a couple dollar's worth of PVC components. Although PVC silencer components aren't as straight and true as machined metal units, they'll do for simple, "forgiving" designs that don't need to be perfectly aligned—like those that use corrugated cardboard washers as baffles. They'll also do for even simpler "lined expansion chamber" type designs; just cut a section of fiberglass or foam rubber pipe insulation and slide it into the tube.

PVC #1

Materials needed:
- 1 1/4″ PVC tubing
- 1″ x 3/4″ PVC bushing
- 1″ x 1/2″ PVC bushing
  (unthreaded)
- top bibb washer

Insert washer into unthreaded bushing; glue if necessary. (Washer makes exit hole smaller and doesn't let quite as much noise escape.)

Wrap masking tape around bushings to get tight fit in tube.

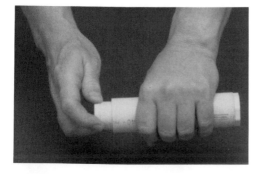

Load tube with silencer "guts" of your choice. If you want to include a wipe, use bushing to trace circumference of foam rubber disk. Friction fit bushings into end of tube.

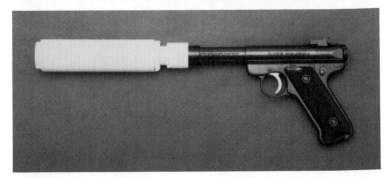

Install on weapon.

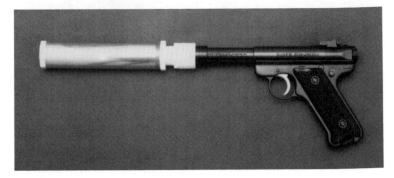

A variant of this design can be made with a section of brass plumbing fixture tubing. The bushings fit this particular tubing so snugly that just one wrap of tape makes them too big to fit back into the tubing. Instead of wrapping them with tape, spray them with layers of primer until they fit tightly.

PVC #2

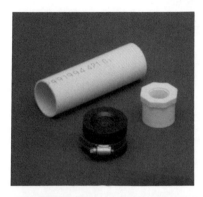

Materials needed:
- 1 1/2″ PVC tubing
  (thin wall)
- 1 1/2″ pipe cap
- 1 1/4″ x 3/4″ PVC bushing

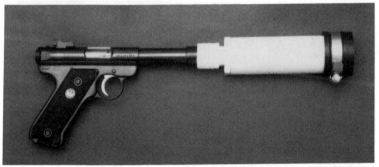

This silencer tube is basically the same as #1 except that it is larger in diameter and has a rubber cap on the end instead of a PVC bushing. (The cap also acts as a wipe.) The particular kind of PVC stock used for this tube has thinner walls and a greater inside diameter than regular 1 1/2″ PVC tubing. Although the 1 1/4″ bushing won't fit inside regular 1 1/2″ PVC tubing, it will fit inside this thinner-walled variety.

SPRINKLER RISER SILENCER

Materials needed:
- spring-loaded riser
- section of soft foam rubber insulation for 1/2″ pipe
- 1″ x 3/4″ PVC bushing
- 3/4" crutch tip

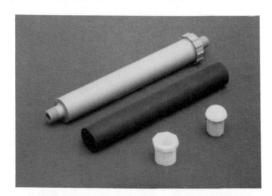

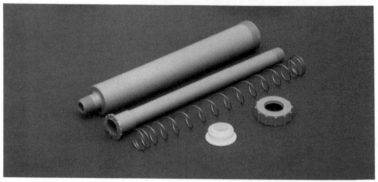

A spring-loaded riser connects a sprinkler head to a buried water line. When the water to the line is turned on, the riser pushes the sprinkler head up above the flora being irrigated. When the water is cut off, the riser lowers the sprinkler head back to its original, less conspicuous position. This riser consists of an outer housing, an inner flanged tube, a spring, a screw-on cap, and a cap liner. The spring and the cap liner may be discarded.

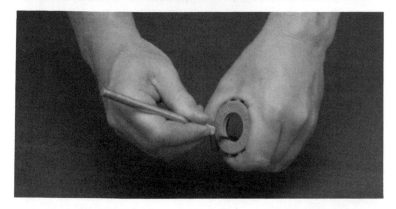

Shave inside diameter of cap to make it big enough to fit over threaded part of coupling on end of weapon.

The inside of the housing has three "fins" that run the length of the housing. They fit into slots in flange of inner tube to keep it from rotating inside housing. These fins need to be cut back flush to inside wall of housing for a distance of about 1 1/2″ so bushing can be inserted into mouth of housing. For best results, use "chisel" type blade in modeling knife. Cut with beveled side of chisel edge turned toward wall of housing.

Cut pipe insulation into short sections and pack them into housing with flanged tube. Leave just enough room in mouth of housing to insert bushing snugly against last segment of insulation.

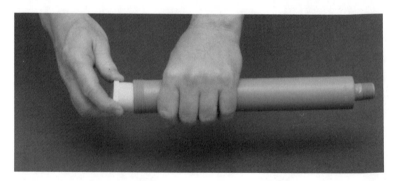

Wrap bushing with tape and friction fit it into housing. Screw cap down over bushing to hold it in place.

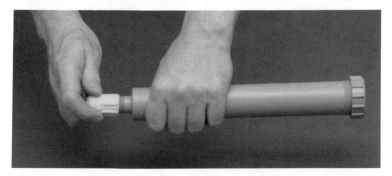

Fit crutch tip onto other end of riser housing. Crutch tip functions as wipe.

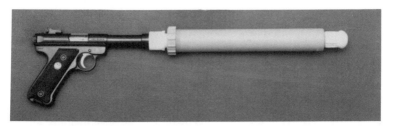

Install on weapon.

COUPLING #2

This coupling is made from a 1″ x 3/4″ PVC reducing male adapter friction fitted around the barrel with the threaded portion situated somewhere behind the muzzle instead of just in front of it. This allows the gun barrel to be inserted right up into the silencer, shortening the overall length of the silenced weapon and eliminating the narrow isthmus connecting the silencer to the barrel.

It will work with any previous design that uses a bushing larger than 1″ x 3/4″. Simply substitute a corresponding bushing threaded for 1″ stock instead of 3/4″ stock.

The inside diameter of this particular reducing male adapter is smaller in the threaded half and slightly larger in the unthreaded half. Two separate, side-by-side 1 1/2″ tape wraps have to be made on the barrel to fit the two different inside diameters.

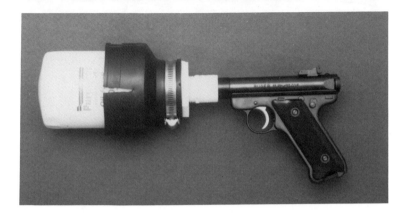

Shown here is the oil filter #1 design mounted on this coupling. Compare with the same unit mounted on coupling #1 on page 33.

CAVEAT

The next section of this book will explore silencer designs for larger centerfire weapons: the Colt .45 and the SKS carbine. These guns generate considerably more internal pressure and fire much larger projectiles than a .22. Therefore, certain safety precautions must be taken.

First of all, make sure everything is either clamped and/or glued together solidly. Don't use friction fits or omit band clamps.

Second, punch holes in the ends of the oil filters. This is a good way to make sure that too much pressure doesn't build up inside and blow the silencer apart.

Third, pay close attention to alignment. That .30 or .45 caliber bullet needs to go through those little threaded holes in the bottoms of those oil filters. If the alignment is off, the bullet will strike the rim of the threaded hole with unpleasant results.

And finally, be careful, use common sense, and always wear eye protection.

Weapon #2
Colt .45

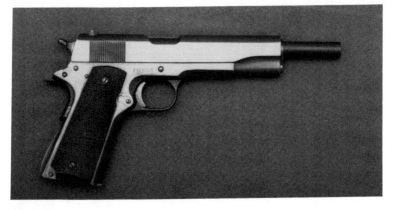

The venerable Colt .45 semiautomatic pistol has been around since 1911 and has a well-earned reputation for toughness and reliability. It fires a large subsonic bullet, and its .45 ACP ammunition is also used in the silenced MAC-10 submachine gun. The Colt .45 fieldstrips very easily, and barrels can be changed in just a few minutes with no tools required. Replacing the regular 5 inch barrel with a 7 inch barrel provides a couple inches of exposed barrel on which to mount a coupling for a silencer.

COUPLING #3

Cut 2″ section from end of 3/4″ PVC nipple. Make four evenly spaced, longitudinal slots that extend down to the threading.

Wrap exposed portion of .45 barrel with 2″ metal repair tape.

After friction fitting PVC section onto taped barrel, secure it *tightly* with two small band clamps. (If you think you need a little more "grab" in the friction fit, try dabbing contact surfaces with a weak glue or any sticky, tacky substance prior to clamping. Don't use too much glue or too strong a glue, though, or you'll end up having to cut the PVC off the barrel.)

OIL FILTER #4

This is essentially the same design as shown on page 33. Note, however, that the oil filter is tightly clamped into the adapter. For extra strength, it can be glued in prior to clamping. Also, although it is not visible, the end of the oil filter has a hole punched in its center. Try filling the filter with water and draining it to wet filter element inside. This might help cool the muzzle blast gases and dampen the noise even more.

CLOG BUSTER #2

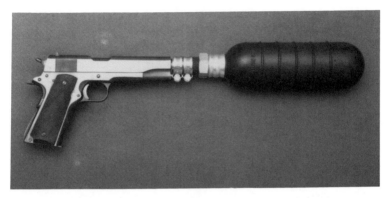

This is simply a larger version of the device shown on page 30. It has eight times the volume of its smaller counterpart and a larger, sturdier hose attachment.

Weapon #3
SKS Carbine

T here are probably more SKS carbines in the world than there are any other particular firearm. Produced in the former Soviet Union and mainland China, the SKS was the predecessor of the infamous AK-47 and uses the same 7.62x39mm ammunition. As of this writing, some models of the SKS carbine are selling for well under $100.

A great variety of accouterments are available for the SKS. The model shown here sports a flash suppressor, perforated steel handguard, scope with "see through" mount, and rubber buttpad. The flash suppressor can be converted into a silencer coupling (as can a muzzle brake). Since disposable silencers for the SKS can be rather bulky, an elevated scope is necessary to get an unobstructed view of the target. See-through scope mounts elevate the scope so the shooter can see underneath it and still use the gun's original iron sights.

COUPLING #4

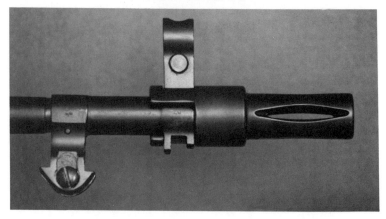

This coupling can be constructed from a flash suppressor (shown here) or a muzzle brake.

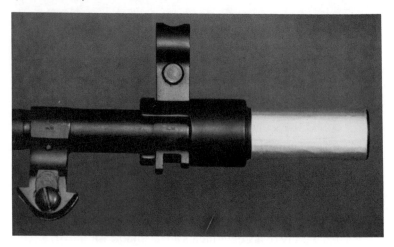

Wrap end of flash suppressor with metal repair tape.

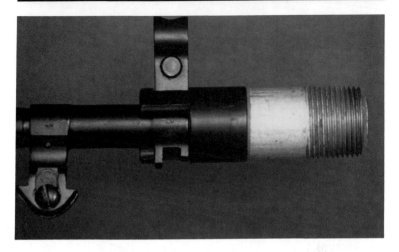

Cut 1 3/4″ section from end of 3/4″ galvanized nipple and make sure it fits snugly on taped portion of flash suppressor. Epoxy it on.

Although some flash suppressors or muzzle brakes may fit snugly on the end of the gun, many have a little "play" in them. A little play at the end of the barrel, however, translates into a lot of play at the end of a long, heavy silencer. The silencer will wobble around and go in and out of alignment if the coupling isn't attached tightly to the barrel.

To remedy this problem, spray a thin coat of primer into the back end of the coupling, being sure to cover all the areas that come in contact with the gun barrel. Let the primer dry, try the coupling once more, and repeat the process until you get a tight fit.

OIL FILTER #5

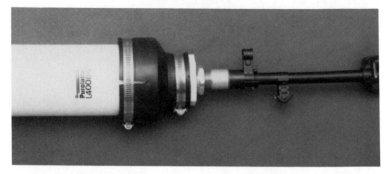

This is pretty much the same design as oil filter #1, except that it uses a longer 3 3/4″ diameter oil filter and has a sturdier bushing assembly. The oil filter has a hole punched in its end and is clamped and glued into adapter.

Instead of gluing 1″ x 3/4″ PVC bushing into unthreaded 2″ x 1″ PVC bushing, screw 1″ x 3/4″ galvanized bushing into threaded 2″ x 1″ PVC bushing.

OIL FILTER #6

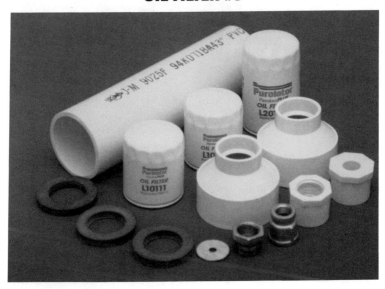

Materials needed:
- 1′ section of 3″ PVC pipe
- 4 3/4″ x 3″ oil filter
- 3 3/8″ x 3″ oil filters (2)
- 3″ x 1 1/2″ PVC reducing couplings (2)
- 1 1/2″ x 1/2″ PVC bushing (unthreaded)
- 1 1/2″ x 1″ PVC bushing
- 3/4″ x 1/2″ galvanized reducing coupling
- 1″ x 3/4″ galvanized bushing
- waste overflow washers, 3/8″ thickness (3)
- top bibb washer

Screw 1″ x 3/4″ galvanized bushing into 1 1/2″ x 1″ PVC bushing. Glue for extra strength.

Glue 3/4″ x 1/2″ galvanized reducing coupling into the other end of 1 1/2″ x 1″ PVC bushing. Fill gap between outside of coupling and inside of bushing with Liquid Steel.

Glue bushing into 3″ x 1 1/2″ reducing coupling.

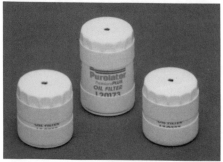

Punch holes in tops of oil filters and wrap with masking tape. Wrap should be just enough to keep them from sliding around or "free falling" when inserted into pipe.

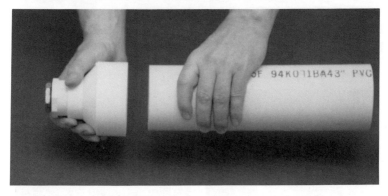

Glue coupling/bushing assembly onto one end of pipe.

Glue top bibb washer into 1 1/2″ x 1/2″ PVC bushing.

Glue bushing with bibb washer into other 3″ x 1 1/2″ reducing coupling.

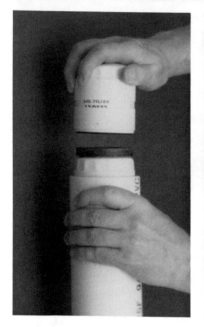 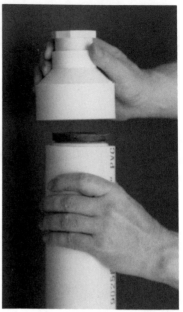

Stand pipe on end and insert oil filters bottom first, beginning with 4 3/4″ oil filter. Place waste overflow washers between oil filters and on top of last filter. Push column of filters all the way down to bottom of pipe.

Glue other 3″ x 1 1/2″ reducing coupling onto open end of pipe. Push it all the way down against waste overflow washer on top of last filter.

Install on weapon.

Bibliography

BOOKS

Anderson, Keith. *How to Build Practical Firearm Suppressors*. Miami, FL: J. Flores Publications, 1994.

Firearm Silencers, Vol. III. El Dorado, AR: Desert Publications, 1975.

Flores, J. *How to Make Disposable Silencers*. Miami, FL: J. Flores Publications, 1984.

_____. *How to Make Disposable Silencers, Vol. II*. Miami, FL: J. Flores Publications, 1985.

Hayduke, George. *The Hayduke Silencer Book: Quick and Dirty Homemade Silencers*. Boulder, CO: Paladin Press, 1989.

_____. *Silent but Deadly: More Homemade Silencers from Hayduke the Master*. Boulder, CO: Paladin Press, 1995.

Home Workshop Silencers I. Boulder, CO: Paladin Press, 1980.

How to Make a Silencer for a .45. Boulder, CO: Paladin Press, 1995.

How to Make a Silencer for a .22. Boulder, CO: Paladin Press, 1994.

Huebner, Siegfried F. *Silencers for Hand Firearms*. Boulder, CO: Paladin Press, 1976.

Minnery, John A. *Firearm Silencers, Vol. II*. El Dorado, AR: Desert Publications, 1981.

Skochko, Leonard and Harry A. Greveris. *Silencers*. Philadelphia: Frankford Arsenal Report No. R-1896, 1968.

Truby, J. David. *Modern Firearm Silencers: Great Designs, Great Designers*. Boulder, CO: Paladin Press, 1992.

_____. *Quiet Killers*. Boulder, CO: Paladin Press, 1992.

_____. *Silencers, Snipers, and Assassins*. Boulder, CO: Paladin Press, 1972.

Wilson, Nolan. *Firearm Silencers*. El Dorado, AR: Desert Publications, 1983.

_____. *The Silencer Cookbook: .22 Rimfire Silencers*. El Dorado, AR: Desert Publications, 1983.

BOOKLET SERIES

Silencers: .223. Hurst, TX: Minuteman Publications, 1985.
Suppressors (six volumes). Hurst, TX: Minuteman Publications, 1982-1983.
Vol. I, Ruger pistol, 1982.
Vol. II, Ruger 10/22, 1982.
Vol. III, AR-7, 1983.

Vol. IV, Uzi, 1983.
Vol. V, MAC 10/11, 1983.
Vol. VI, KG-9 & KG-99, 1983.

VIDEOS

Deadly Weapons: Firearms & Firepower. Pinole, CA: Anite Productions, 1984.

Improvised Suppressors: Secrets of Silencing Firearms. Boulder, CO: Paladin Press, 1994.

Whispering Death: Secrets of Improvised and State-of-the-Art Silencers. Boulder, CO: Paladin Press, 1989.